● WHY THIS BOOK MATTERS

Earth is our home, a planet full of breathtaking landscapes, diverse wildlife, and natural wonders. But as our world faces environmental challenges like climate change, deforestation, and pollution, it's more important than ever to understand and protect it. That's why this book was created—to educate, inspire, and engage people in a fun and interactive way!

Inside, you'll find 200 fascinating facts that uncover the beauty and complexity of our planet, followed by 200 trivia questions that challenge your knowledge and spark curiosity. Whether you're a teacher, a student, an eco-enthusiast, or just someone looking to learn more, this book is designed for you.

So, are you ready to test your environmental knowledge? Let's dive in and celebrate Earth Day not just once a year, but every day!

HISTORY OF EARTH DAY

- Earth Day was first celebrated on April 22, 1970.
- It was founded by U.S. Senator Gaylord Nelson to raise awareness about environmental issues.
- The first Earth Day led to the creation of the Environmental Protection Agency (EPA) in the U.S.
- More than 20 million Americans participated in the first Earth Day events.
- Earth Day became a global event in 1990, involving over 140 countries.

- April 22 was chosen to maximize student participation, as it falls between spring break and final exams.
- The Earth Day Network coordinates events worldwide.
- The Paris Climate Agreement was signed on Earth Day 2016.
- Earth Day is now recognized as the largest secular observance in the world.
- The theme for Earth Day changes every year to address different environmental challenges.

GENERAL ENVIRONMENTAL FACTS

- The Earth is around 4.5 billion years old.
- More than 70% of the Earth's surface is covered by water.
- The Amazon Rainforest produces 20% of the world's oxygen.
- Coral reefs are often called the "rainforests of the sea."
- The Great Pacific Garbage Patch is twice the size of Texas.
- Antarctica is the coldest, driest, and windiest continent

- Deforestation leads to the loss of 18 million acres of forest each year.
- Every second, a section of rainforest the size of a football field is destroyed.
- The world's population produces about 2.01 billion metric tons of solid waste annually.
- Plastic pollution kills over 1 million marine animals every year.

CLIMATE CHANGE AND GLOBAL WARMING

- The last decade was the hottest on record.
- Carbon dioxide (CO_2) levels are the highest they have been in 800,000 years.
- The Arctic is warming twice as fast as the global average.
- Rising sea levels threaten coastal cities worldwide.
- Climate change has intensified natural disasters such as hurricanes and wildfires.
- Greenhouse gases trap heat in the Earth's atmosphere.

- The burning of fossil fuels is the primary cause of global warming.
- Methane is 25 times more effective at trapping heat than carbon dioxide.
- Ocean acidification is caused by increased CO_2 absorption in seawater.
- Renewable energy sources like wind and solar can reduce greenhouse gas emissions.

SUSTAINABILITY AND CONSERVATION

- Recycling one aluminum can saves enough energy to power a TV for three hours.
- Glass takes over 4,000 years to decompose in landfills.
- Bamboo is one of the most sustainable plants due to its rapid growth.
- Composting reduces methane emissions from landfills.
- LED light bulbs use 75% less energy than traditional bulbs.
- A single tree can absorb up to 48 pounds of CO_2 per year.

- Turning off electronics when not in use can save significant amounts of energy.
- Fast fashion is a major contributor to environmental pollution.
- The average person generates over 4 pounds of trash per day.
- Buying local food reduces carbon emissions from transportation.

WILDLIFE AND BIODIVERSITY

- One million species are at risk of extinction due to human activities.
- Bees are responsible for pollinating about 75% of all crops.
- Rainforests contain more than half of the world's species.
- Elephants help create water holes that benefit other animals.
- The ocean produces over 50% of the world's oxygen.
- Some species of sea turtles have existed for over 100 million years.

- The polar bear population is declining due to melting ice caps.
- Sharks have been around longer than dinosaurs.
- Deforestation is the biggest threat to orangutans.
- Coral bleaching occurs when ocean temperatures rise too high.

POLLUTION AND WASTE

- Around 8 million metric tons of plastic enter the ocean each year.
- Air pollution causes about 7 million deaths annually.
- The average plastic bag is used for 12 minutes but takes centuries to decompose.
- Cigarette butts are the most common form of litter.
- Oil spills can cause devastating damage to marine life.
- Microplastics have been found in the deepest parts of the ocean.

- Landfills are one of the largest sources of methane emissions.
- E-waste is the fastest-growing type of waste in the world.
- A single plastic straw can take up to 200 years to break down.
- Noise pollution affects both humans and animals.

WATER CONSERVATION

- Less than 1% of the Earth's water is drinkable.
- A running faucet can waste up to 2.5 gallons of water per minute.
- Fixing a leaking tap can save up to 3,000 gallons of water per year.
- The average person uses about 80-100 gallons of water daily.
- A five-minute shower uses around 10-25 gallons of water.
- It takes about 1,800 gallons of water to produce one pound of beef.

- Desalination is becoming a popular method of producing fresh water.
- Droughts are becoming more frequent due to climate change.
- Wetlands help filter and clean water naturally.
- Water pollution can lead to the spread of diseases like cholera.

RENEWABLE ENERGY

- Solar energy is the fastest-growing renewable energy source.
- Wind turbines can generate electricity without producing pollution.
- Hydropower provides about 16% of the world's electricity.
- Geothermal energy uses heat from the Earth's core to generate power.
- Biofuels are made from organic materials like plants and algae.
- The sun produces enough energy in one hour to power the entire world for a year.

- The largest solar power plant is located in India.
- Wind energy capacity has doubled in the past decade.
- Hydrogen fuel cells are an emerging clean energy technology.
- Some countries aim to be carbon neutral by 2050.

ECO-FRIENDLY HABITS

- Bringing reusable bags reduces plastic waste.
- Carpooling reduces carbon emissions.
- Eating less meat can lower your carbon footprint.
- Turning off lights saves energy.
- Planting trees helps combat climate change.
- Riding a bike instead of driving reduces air pollution.
- Using public transport decreases overall fuel consumption.
- Second-hand shopping reduces waste.

- Using a refillable water bottle helps reduce plastic waste.
- Supporting eco-friendly brands encourages sustainability.

ECO-FRIENDLY INNOVATIONS

- Some companies are developing biodegradable plastics made from plants.
- Electric vehicles (EVs) produce zero tailpipe emissions.
- Smart thermostats help reduce energy consumption.
- Vertical farming uses less land and water than traditional agriculture.
- Carbon capture technology helps remove CO_2 from the air.
- Some cities are testing roads made from recycled plastic.

- Algae-based biofuels could become a sustainable alternative to gasoline.
- 3D-printed houses made from recycled materials are gaining popularity.
- Floating solar farms are being built on reservoirs to save land space.
- Some clothing brands are using recycled ocean plastics to make fabric.

EARTH'S NATURAL WONDERS

- The Amazon Rainforest is home to 10% of all known species.
- The Great Barrier Reef is the largest coral reef system in the world.
- The Mariana Trench is the deepest part of the ocean.
- Iceland generates nearly 100% of its electricity from renewable sources.
- The Sahara Desert was once a lush green landscape.
- The Northern Lights (Aurora Borealis) are caused by charged particles from the sun.

- The Dead Sea is one of the saltiest bodies of water on Earth.
- Lake Baikal in Russia holds about 20% of the world's unfrozen freshwater.
- The Grand Canyon was carved by the Colorado River over millions of years.
- Some trees, like the Methuselah tree, are over 4,800 years old.

ENDANGERED SPECIES AND CONSERVATION EFFORTS

- The IUCN Red List tracks endangered and extinct species.
- Habitat destruction is the main cause of species extinction.
- The giant panda was removed from the endangered species list in 2016.
- The Tasmanian tiger went extinct in 1936.
- Some rhino species are critically endangered due to poaching.

- Conservation programs have helped increase the bald eagle population.
- The blue whale is the largest animal ever to exist.
- Tigers have lost 93% of their historical range.
- Polar bears rely on sea ice for hunting and survival.
- Gorillas share about 98% of their DNA with humans.

OCEANS AND MARINE LIFE

- The ocean covers about 71% of the Earth's surface.
- Over 90% of the ocean remains unexplored.
- Mangroves help protect coastlines from erosion and storms.
- The Great Pacific Garbage Patch contains an estimated 1.8 trillion pieces of plastic.
- Some jellyfish have been around for over 500 million years.

- Ocean currents help regulate global temperatures.
- Sea turtles can live over 100 years.
- Coral reefs support about 25% of all marine species.
- Plastic pollution is found even in the deepest parts of the ocean.
- Whales play a key role in the marine ecosystem by cycling nutrients.

DEFORESTATION AND REFORESTATION

- The Amazon Rainforest loses thousands of acres of trees daily.
- Reforestation projects help restore lost forests.
- Trees absorb carbon dioxide and release oxygen.
- Some tree species can live for thousands of years.
- Urban forests help improve air quality in cities.
- Forests cover about 31% of the Earth's land area.

- Some companies are using drones to plant trees more efficiently.
- Agroforestry combines agriculture and tree planting to improve sustainability.
- Deforestation contributes to climate change by releasing stored carbon.
- Rainforests are disappearing at an alarming rate due to logging and agriculture.

SUSTAINABLE AGRICULTURE AND FOOD WASTE

- Organic farming avoids the use of synthetic pesticides and fertilizers.
- Crop rotation helps maintain soil fertility.
- Bees and butterflies are essential for pollination.
- Food waste accounts for about one-third of all food produced.
- Growing your own food reduces your carbon footprint.
- Buying seasonal produce supports local farmers and reduces emissions.

- Some restaurants are using food scraps to create new dishes.
- Vertical farms can grow crops in urban areas with less water.
- Precision agriculture uses technology to optimize farming efficiency.
- Reducing meat consumption can significantly lower greenhouse gas emissions.

ECO-FRIENDLY TECHNOLOGIES

- Solar panels work even on cloudy days.
- Wind energy is one of the fastest-growing renewable energy sources.
- Hydroelectric power plants use flowing water to generate electricity.
- Geothermal energy harnesses heat from beneath the Earth's surface.
- Some countries generate over 90% of their energy from renewables.

- Hydrogen fuel cells produce electricity with only water as a byproduct.
- Smart homes can reduce energy waste.
- Electric buses are being introduced in many cities to reduce pollution.
- Some batteries are now recyclable and less harmful to the environment.
- Innovations in energy storage are helping make renewable energy more efficient.

EXTREME WEATHER AND NATURAL DISASTERS

- limate change has increased the intensity of hurricanes.
- Wildfires are more common due to rising global temperatures.
- Floods are becoming more frequent in many parts of the world.
- Droughts affect millions of people every year.
- Tornado Alley in the U.S. experiences the highest number of tornadoes.

- Some islands are at risk of disappearing due to rising sea levels.
- Volcanic eruptions can impact global climate.
- Earthquakes occur due to movement of tectonic plates.
- Some regions are experiencing longer and more intense heat waves.
- Natural disasters often displace large populations.

INTERESTING EARTH FACTS

- Earth is the only known planet with liquid water on its surface.
- The ozone layer protects us from harmful UV radiation.
- The Earth's magnetic field shields us from solar winds.
- The Moon affects ocean tides on Earth.
- The longest river in the world is the Nile.
- The Sahara Desert is expanding due to climate change.

- The highest mountain on Earth is Mount Everest.
- The Earth's rotation is gradually slowing down.
- There are more than 1,500 potentially active volcanoes worldwide.
- Lightning strikes the Earth about 100 times per second.

HOW INDIVIDUALS CAN MAKE A DIFFERENCE

- Using reusable water bottles reduces plastic waste.
- Turning off lights when leaving a room saves energy.
- Walking or biking instead of driving cuts carbon emissions.
- Supporting sustainable brands helps the environment.
- Eating plant-based meals lowers your environmental footprint.
- Spreading awareness about climate change encourages action.

- Buying second-hand clothes reduces waste.
- Reducing single-use plastics helps prevent ocean pollution.
- Participating in tree-planting events benefits the planet.
- Writing to policymakers can promote environmental laws.

SURPRISING ENVIRONMENTAL FACTS

- The world's largest iceberg was bigger than Jamaica.
- Some fish can change their gender due to environmental changes.
- A single reusable bag can replace thousands of plastic bags.
- Some plants can absorb toxins from the air and soil.
- The cleanest air in the world is found in Antarctica.
- The world's oldest tree is over 9,500 years old.

- Earth experiences about 500,000 earthquakes each year.
- The first Earth Day led to the banning of harmful pesticides like DDT.
- Switching to a plant-based diet can reduce water use by 50%.
- Scientists estimate we have less than 10 years to significantly reduce carbon emissions.

200 MULTIPLE-CHOICE TRIVIA QUESTIONS

HISTORY OF EARTH DAY

- When was the first Earth Day celebrated?
- a) 1965
- b) 1970 ■
- c) 1985
- d) 1990
- Who is credited with founding Earth Day?
- a) Al Gore
- b) Rachel Carson
- c) Gaylord Nelson ■
- d) Jane Goodall
- What major environmental agency was created in the U.S. after the first Earth Day?
- a) EPA ■
- b) WHO
- c) UNESCO
- d) Greenpeace

- How many people participated in the first Earth Day?
- a) 1 million
- b) 10 million
- c) 20 million ■
- d) 50 million
- In which year did Earth Day become a global event?
- a) 1970
- b) 1985
- c) 1990 ■
- d) 2000

CLIMATE CHANGE & GLOBAL WARMING

What is the main cause of climate change?
- a) Volcanoes
- b) Greenhouse gases ■
- c) The ozone layer
- d) Solar flares

Which gas is the biggest contributor to global warming?
- a) Oxygen
- b) Nitrogen
- c) Carbon dioxide ■
- d) Hydrogen

What effect does melting ice caps have on the planet?
- a) Rising sea levels ■
- b) More rainfall
- c) Stronger earthquakes
- d) Less wind

- The Arctic is warming at what rate compared to the rest of the planet?
- a) The same rate
- b) Twice as fast ■
- c) Half as fast
- d) It is not warming
- What was the warmest decade on record?
- a) 1980s
- b) 1990s
- c) 2000s
- d) 2010s ■

POLLUTION & WASTE

- What is the largest source of plastic waste in the ocean?
- a) Fishing nets
- b) Plastic bags
- c) Bottles
- d) Single-use plastics ■
- How long does it take for a plastic bottle to decompose?
- a) 10 years
- b) 100 years
- c) 450 years ■
- d) 1,000 years
- What is the most common type of litter found in the ocean?
- a) Plastic straws
- b) Cigarette butts ■
- c) Aluminum cans
- d) Glass bottles

- What percentage of plastic waste is recycled worldwide?
- a) 9% ■
- b) 25%
- c) 50%
- d) 75%
- What is the term for small plastic particles found in the ocean?
- a) Microfibers
- b) Microplastics ■
- c) Polymers
- d) Nano-waste

RENEWABLE ENERGY

What is the most abundant renewable energy source on Earth?
- a) Wind
- b) Hydropower
- c) Solar ■
- d) Biomass

What type of energy is generated by the movement of air?
- a) Geothermal
- b) Wind ■
- c) Hydroelectric
- d) Nuclear

Which country generates the most wind power?
- a) China ■
- b) USA
- c) Germany
- d) India

Which renewable energy source relies on the heat from the Earth's core?
- a) Hydropower
- b) Geothermal ■
- c) Biomass
- d) Solar

What is the main disadvantage of solar power?
- a) It is expensive
- b) It doesn't work at night ■
- c) It causes pollution
- d) It takes up too much space

WILDLIFE & BIODIVERSITY

What animal is often considered an indicator of climate change?
- a) Penguin
- b) Polar bear ■
- c) Elephant
- d) Tiger

What percentage of the world's species live in rainforests?
- a) 10%
- b) 25%
- c) 50% ■
- d) 75%

What is the main cause of habitat destruction?
- a) Earthquakes
- b) Agriculture ■
- c) Hunting
- d) Tourism

Which species is critically endangered due to poaching for its horns?
- a) Elephant
- b) Rhino ■
- c) Giraffe
- d) Dolphin

How much of the ocean has been explored by humans?
- a) 5% ■
- b) 20%
- c) 50%
- d) 75%

ECO-FRIENDLY HABITS

What is the best way to reduce plastic waste?
- a) Burn it
- b) Recycle it
- c) Avoid using it ■
- d) Throw it in the ocean

What is the most eco-friendly way to travel?
- a) Carpooling
- b) Taking the bus
- c) Walking ■
- d) Driving an electric car

What is a zero-waste lifestyle?
- a) Not producing any garbage ■
- b) Recycling everything
- c) Using only biodegradable materials
- d) Reducing food waste

Why is eating local food better for the environment?
- a) It tastes better
- b) It supports the local economy
- c) It reduces carbon emissions ■
- d) It is cheaper

What is the best alternative to single-use plastic straws?
- a) Glass straws ■
- b) Paper straws
- c) Metal straws
- d) Bamboo straws

DEFORESTATION & REFORESTATION

What percentage of the Earth's forests have been destroyed?
- a) 10%
- b) 25%
- c) 50% ■
- d) 75%

What is the leading cause of deforestation?
- a) Urban development
- b) Logging
- c) Agriculture ■
- d) Wildfires

Which of the following is a benefit of planting trees?
- a) Reducing carbon dioxide in the air ■
- b) Increasing soil erosion
- c) Producing more pollution
- d) Making the land dry

What is the term for planting trees to restore a forest?
- a) Deforestation
- b) Afforestation
- c) Reforestation ■
- d) Carbon sequestration

How do trees help reduce global warming?
- a) By absorbing methane
- b) By absorbing carbon dioxide ■
- c) By releasing oxygen
- d) By increasing temperatures

OCEANS & MARINE LIFE

What is the largest ocean on Earth?
- a) Atlantic Ocean
- b) Indian Ocean
- c) Pacific Ocean ■
- d) Arctic Ocean

What percentage of the Earth's surface is covered by oceans?
- a) 50%
- b) 60%
- c) 71% ■
- d) 90%

What is the biggest threat to coral reefs?
- a) Hurricanes
- b) Pollution
- c) Rising ocean temperatures ■
- d) Overfishing

What is "bycatch"?
- a) Fishing in deep waters
- b) Catching unwanted marine animals ■
- c) A type of pollution
- d) Illegal fishing

What is the name of the massive garbage patch floating in the Pacific Ocean?
- a) Atlantic Plastic Patch
- b) Indian Ocean Waste Zone
- c) Great Pacific Garbage Patch ■
- d) Arctic Debris Field

SUSTAINABLE LIVING

What is the best way to reduce energy use at home?
- a) Using LED bulbs ■
- b) Keeping appliances on standby
- c) Using more air conditioning
- d) Keeping the lights on all day

What is an eco-friendly alternative to traditional plastic?
- a) Biodegradable plastic ■
- b) Styrofoam
- c) Aluminum
- d) Glass

What is "fast fashion"?
- a) Clothing made with recycled materials
- b) Cheap, mass-produced clothing ■
- c) Clothes that are made to last
- d) Sustainable fashion brands

What is one way to reduce food waste?
- a) Throwing away leftovers
- b) Buying more food than needed
- c) Composting food scraps ■
- d) Keeping expired food

Which of the following actions helps conserve water?
- a) Taking long showers
- b) Running the tap while brushing teeth
- c) Fixing leaks and using low-flow faucets ■
- d) Washing clothes daily

EXTREME WEATHER & NATURAL DISASTERS

What is the strongest type of storm?
- a) Tornado
- b) Thunderstorm
- c) Hurricane ■
- d) Dust storm

What is the main cause of rising sea levels?
- a) Earthquakes
- b) Melting ice caps ■
- c) Volcanoes
- d) Rainfall

What is a drought?
- a) An area with too much water
- b) A long period of little or no rainfall ■
- c) A severe storm
- d) A type of earthquake

Which area is most affected by hurricanes?
- a) The Sahara Desert
- b) The Arctic
- c) Coastal regions ■
- d) Mountain areas

What is the main cause of wildfires?
- a) Lightning
- b) Human activity ■
- c) Heavy rainfall
- d) Earthquakes

INTERESTING EARTH FACTS

What is the longest river in the world?
- a) Amazon River
- b) Nile River ■
- c) Yangtze River
- d) Mississippi River

How old is the Earth?
- a) 2 billion years
- b) 4.5 billion years ■
- c) 6 billion years
- d) 10 billion years

What is the only planet known to support life?
- a) Mars
- b) Jupiter
- c) Venus
- d) Earth ■

Which layer of the atmosphere protects us from harmful UV rays?
- a) Troposphere
- b) Ozone layer ■
- c) Stratosphere
- d) Thermosphere

What percentage of the Earth's water is freshwater?
- a) 3% ■
- b) 10%
- c) 25%
- d) 50%

SUSTAINABLE AGRICULTURE & FOOD PRODUCTION

What is the term for growing food without synthetic pesticides or fertilizers?
- a) GMO farming
- b) Organic farming ■
- c) Industrial agriculture
- d) Conventional farming

Which farming method helps conserve water the most?
- a) Flood irrigation
- b) Drip irrigation ■
- c) Overhead sprinklers
- d) Rain-fed farming

What is a plant-based diet's main environmental benefit?
- a) It requires less land and water ■
- b) It provides more protein
- c) It reduces fruit production
- d) It increases carbon emissions

What is the term for growing food in urban areas?
- a) Urban forestry
- b) Vertical farming ■
- c) Conventional agriculture
- d) Agroforestry

Which crop requires the most water to grow?
- a) Wheat
- b) Rice ■
- c) Corn
- d) Potatoes

WATER CONSERVATION & POLLUTION

What is the main cause of water pollution?
- a) Natural disasters
- b) Industrial waste ■
- c) Tree planting
- d) Wind energy

What is the process of removing salt from seawater to make it drinkable?
- a) Filtration
- b) Distillation
- c) Desalination ■
- d) Evaporation

What is a major source of freshwater pollution?
- a) Ocean currents
- b) Oil spills
- c) Chemical runoff from farms ■
- d) Solar energy

Which country has the most freshwater resources?
- a) USA
- b) Russia
- c) Brazil
- d) Canada

What is the most efficient way to save water at home?
- a) Taking shorter showers
- b) Watering lawns daily
- c) Washing dishes under running water
- d) Using a lot of detergent

RECYCLING & WASTE MANAGEMENT

What is the "3Rs" principle of waste management?
- a) Reduce, Recycle, Reuse ■
- b) Repair, Recycle, Rebuild
- c) Reuse, Refuse, Replant
- d) Recycle, Refuel, Restore

What is the main challenge with recycling plastic?
- a) It takes up too much space
- b) It is too light
- c) Different types of plastics require different recycling processes ■
- d) It is biodegradable

What material is easiest to recycle?
- a) Plastic
- b) Glass ■
- c) Styrofoam
- d) Batteries

What happens when electronic waste (e-waste) is not properly disposed of?
- a) It releases toxic chemicals ■
- b) It turns into compost
- c) It becomes a renewable energy source
- d) It helps plants grow

Which of the following is NOT commonly recyclable?
- a) Cardboard
- b) Aluminum cans
- c) Plastic bags ■
- d) Newspapers

ALTERNATIVE ENERGY INNOVATIONS

What is the fastest-growing renewable energy source?
- a) Wind energy
- b) Solar energy ■
- c) Biomass
- d) Hydropower

What are biofuels made from?
- a) Rocks
- b) Organic matter like plants and algae ■
- c) Coal
- d) Nuclear waste

What is the most efficient type of renewable energy?
- a) Wind
- b) Solar
- c) Hydropower ■
- d) Biomass

Which of these technologies stores excess renewable energy?
- a) Fuel cells ■
- b) Gas turbines
- c) Diesel generators
- d) Fossil fuel tanks

Which country is a leader in geothermal energy?
- a) Iceland ■
- b) Japan
- c) Brazil
- d) Canada

ECO-FRIENDLY ARCHITECTURE & GREEN CITIES

What is a green roof?
- a) A roof made of solar panels
- b) A roof covered with plants ■
- c) A metal roof
- d) A painted roof

What type of building material is most sustainable?
- a) Concrete
- b) Bamboo ■
- c) Plastic
- d) Asphalt

What is the term for cities designed to be environmentally friendly?
- a) Smart cities
- b) Sustainable cities ■
- c) Carbon cities
- d) Industrial cities

What is the biggest benefit of electric public transportation?
- a) It is faster
- b) It reduces air pollution ■
- c) It creates more waste
- d) It costs more

What is an energy-efficient home feature?
- a) Large glass windows facing the sun ■
- b) Thin walls
- c) High-energy appliances
- d) Uninsulated walls

EARTH'S NATURAL RESOURCES

What is a nonrenewable resource?
- a) Water
- b) Trees
- c) Coal ■
- d) Wind

What is the most commonly used fossil fuel?
- a) Natural gas
- b) Oil
- c) Coal ■
- d) Nuclear energy

What mineral is commonly used in making solar panels?
- a) Copper
- b) Silicon ■
- c) Gold
- d) Iron

What is a major problem with overfishing?
- a) It causes more rain
- b) It disrupts marine ecosystems ■
- c) It increases coral reefs
- d) It helps fish reproduce faster

What is the best way to conserve natural resources?
- a) Use more coal
- b) Cut down more trees
- c) Reduce consumption ■
- d) Use more plastic

ENDANGERED SPECIES & BIODIVERSITY

What is the biggest cause of species extinction today?
- a) Hunting
- b) Climate change
- c) Habitat destruction ■
- d) Overpopulation

What does the term "biodiversity" mean?
- a) The variety of life in an ecosystem ■
- b) The number of plants in a rainforest
- c) The amount of oxygen produced by trees
- d) The study of ocean life

What is the name of the list that tracks endangered species?
- a) World Wildlife Record
- b) The Red List ■
- c) Earth Watch
- d) The Green Book

What is an example of an endangered animal?
- a) African elephant
- b) Bald eagle
- c) Siberian tiger ■
- d) House cat

Which habitat has the highest biodiversity?
- a) Grasslands
- b) Coral reefs ■
- c) Deserts
- d) Tundras

AIR POLLUTION & CLIMATE CHANGE

What is the biggest contributor to air pollution?
- a) Volcanic eruptions
- b) Burning fossil fuels ■
- c) Ocean currents
- d) Wind

What greenhouse gas is most responsible for global warming?
- a) Oxygen
- b) Carbon dioxide ■
- c) Nitrogen
- d) Helium

Which human activity produces the most carbon dioxide?
- a) Agriculture
- b) Transportation ■
- c) Fishing
- d) Tree planting

What is the effect of the greenhouse effect?
- a) Cooling of the Earth
- b) Rising global temperatures ■
- c) Reduction in rainfall
- d) Ozone depletion

What is an example of an alternative to fossil fuels?
- a) Coal
- b) Nuclear energy
- c) Solar energy ■
- d) Natural gas

PLASTIC POLLUTION & WASTE REDUCTION

How long does it take for plastic to decompose?
- a) 1 year
- b) 10 years
- c) 100+ years ■
- d) It never decomposes

What percentage of plastic waste is actually recycled?
- a) 5%
- b) 10% ■
- c) 25%
- d) 50%

What is the best way to reduce plastic pollution?
- a) Banning single-use plastics ■
- b) Burning plastic
- c) Recycling all plastic
- d) Using plastic bags

Which material is a good alternative to plastic?
- a) Styrofoam
- b) Paper ■
- c) More plastic
- d) Rubber

What is "microplastic"?
- a) Small fish
- b) Tiny plastic particles in the environment ■
- c) A type of plastic toy
- d) A recycling method

FORESTS & TREE CONSERVATION

What percentage of the Earth's land is covered by forests?
- a) 10%
- b) 20%
- c) 30% ■
- d) 50%

Which country has the largest rainforest?
- a) Indonesia
- b) Brazil ■
- c) Canada
- d) Russia

What is the term for cutting down trees without replacing them?
- a) Reforestation
- b) Urbanization
- c) Deforestation ■
- d) Afforestation

What is an important role of forests?
- a) Absorbing carbon dioxide ■
- b) Producing oil
- c) Making the land drier
- d) Increasing pollution

What type of tree absorbs the most carbon dioxide?
- a) Mango trees
- b) Pine trees
- c) Oak trees
- d) Mangrove trees ■

EARTH DAY & ENVIRONMENTAL AWARENESS

When is Earth Day celebrated?
- a) April 1
- b) April 22 ■
- c) June 5
- d) December 25

What was the first year Earth Day was celebrated?
- a) 1965
- b) 1970 ■
- c) 1985
- d) 1999

Who founded Earth Day?
- a) John Muir
- b) Greta Thunberg
- c) Gaylord Nelson ■
- d) Elon Musk

What is the main goal of Earth Day?
- a) Celebrate space exploration
- b) Raise awareness about environmental issues ■
- c) Promote new technology
- d) Encourage tourism

Which major event happened on Earth Day 2016?
- a) The first Earth Day was celebrated
- b) The Paris Agreement was signed ■
- c) The Amazon Rainforest was protected
- d) The ozone layer was fully repaired

MISCELLANEOUS ENVIRONMENTAL TRIVIA

What is the largest source of renewable energy in the world?
- a) Wind
- b) Solar
- c) Hydropower ■
- d) Geothermal

What is the world's largest desert?
- a) Gobi
- b) Sahara
- c) Antarctic Desert ■
- d) Kalahari

What gas do plants release into the air?
- a) Carbon dioxide
- b) Oxygen ■
- c) Methane
- d) Nitrogen

What is the largest animal on Earth?
- a) Elephant
- b) Blue whale ■
- c) Giraffe
- d) Polar bear

How much of the world's oxygen is produced by the ocean?
- a) 20%
- b) 30%
- c) 50%
- d) 70% ■

OCEANS & MARINE CONSERVATION

What percentage of the Earth's surface is covered by oceans?
- a) 50%
- b) 60%
- c) 70% ■
- d) 80%

What is the largest ocean on Earth?
- a) Atlantic Ocean
- b) Indian Ocean
- c) Arctic Ocean
- d) Pacific Ocean ■

What is coral bleaching caused by?
- a) Rising ocean temperatures ■
- b) Too many fish
- c) Lack of sunlight
- d) Oil spills

What is the name of the largest coral reef system?
- a) Hawaiian Reefs
- b) Great Barrier Reef ■
- c) Caribbean Reef
- d) Pacific Reef

What is an example of a marine protected area?
- a) National parks
- b) Underwater nature reserves ■
- c) Fishing zones
- d) Offshore drilling sites

RENEWABLE ENERGY & SUSTAINABLE PRACTICES

What is the cleanest source of renewable energy?
- a) Nuclear energy
- b) Solar energy ■
- c) Coal
- d) Natural gas

What country produces the most wind energy?
- a) Germany
- b) China ■
- c) USA
- d) Canada

What is an energy-efficient way to heat homes?
- a) Electric heaters
- b) Burning coal
- c) Geothermal heating ■
- d) Gas-powered furnaces

What is a disadvantage of solar energy?
- a) It is non-renewable
- b) It depends on sunlight ■
- c) It causes pollution
- d) It is dangerous to use

What is a benefit of using electric vehicles?
- a) They reduce air pollution ■
- b) They use more oil
- c) They cost less than all gasoline cars
- d) They increase carbon dioxide levels

SUSTAINABLE LIVING & ECO-FRIENDLY HABITS

What is one way to reduce electricity use?
- a) Keeping lights on
- b) Unplugging devices when not in use ■
- c) Using more appliances
- d) Running the AC at full power

What type of shopping bag is most eco-friendly?
- a) Plastic bags
- b) Paper bags
- c) Reusable cloth bags ■
- d) Styrofoam bags

What is composting?
- a) Throwing away food waste
- b) Burning organic materials
- c) Turning organic waste into soil fertilizer ■
- d) Burying plastic waste

What is the best way to save water at home?
- a) Taking longer showers
- b) Fixing leaks ■
- c) Running dishwashers with half loads
- d) Keeping the tap running

Which of the following household items can be composted?
- a) Plastic wrappers
- b) Eggshells ■
- c) Glass bottles
- d) Aluminum cans

CLIMATE SCIENCE & GLOBAL WARMING

What is the biggest effect of climate change?
- a) Rising global temperatures ■
- b) More trees growing
- c) Less pollution
- d) Decreased carbon emissions

What is the term for extreme weather caused by climate change?
- a) Weather cycles
- b) Climate impact
- c) Climate extremes ■
- d) Natural variation

What is a major impact of melting ice caps?
- a) Rising sea levels ■
- b) Lower temperatures
- c) More freshwater
- d) Decreased rainfall

What type of gas is methane?
- a) A greenhouse gas ■
- b) A cooling agent
- c) A fuel source
- d) A solid material

Which sector contributes most to global greenhouse gas emissions?
- a) Agriculture
- b) Transportation
- c) Energy production ■
- d) Recycling

EARTH & SPACE CONNECTION

What is the only planet known to support life?
- a) Mars
- b) Venus
- c) Earth ■
- d) Jupiter

What is the Earth's outermost layer called?
- a) Mantle
- b) Core
- c) Crust ■
- d) Atmosphere

What is the most abundant gas in Earth's atmosphere?
- a) Oxygen
- b) Carbon dioxide
- c) Nitrogen ■
- d) Helium

What is the force that keeps Earth's atmosphere in place?
- a) Wind
- b) Gravity ■
- c) Sunlight
- d) Magnetism

What is the Earth's natural satellite?
- a) Sun
- b) Mars
- c) Moon ■
- d) Comet

ECO-FRIENDLY TECHNOLOGY & INNOVATIONS

What is an example of biodegradable packaging?
- a) Plastic wrap
- b) Cardboard ■
- c) Aluminum foil
- d) Styrofoam

What does an electric grid do?
- a) Stores water
- b) Distributes electricity ■
- c) Cleans air pollution
- d) Creates gasoline

What type of building design helps save energy?
- a) Passive solar design ■
- b) Open rooftops
- c) No windows
- d) Uninsulated walls

What is the function of carbon capture technology?
- a) Increases carbon in the air
- b) Reduces carbon emissions ■
- c) Generates nuclear power
- d) Creates more oil

What is an alternative to traditional gasoline for cars?
- a) Coal
- b) Hydrogen fuel ■
- c) Diesel
- d) Plastic waste

WILDLIFE & ECOSYSTEMS

What is the largest rainforest in the world?
- a) Amazon Rainforest ■
- b) Congo Rainforest
- c) Borneo Rainforest
- d) Daintree Rainforest

Which animal is known as an "ecosystem engineer"?
- a) Elephant
- b) Beaver ■
- c) Lion
- d) Dolphin

What do pollinators like bees help with?
- a) Producing honey
- b) Breaking down waste
- c) Helping plants reproduce ■
- d) Catching insects

Which ecosystem stores the most carbon?
- a) Deserts
- b) Grasslands
- c) Forests
- d) Wetlands ■

What happens if a species goes extinct?
- a) It can easily come back
- b) It disrupts the ecosystem ■
- c) It has no effect on nature
- d) It benefits other species

FINAL QUESTIONS ON ENVIRONMENTAL AWARENESS

What is the best way to protect endangered animals?
- a) Build more zoos
- b) Protect their natural habitats ■
- c) Hunt them responsibly
- d) Keep them as pets

What is the leading cause of wildfires?
- a) Lightning
- b) Human activity ■
- c) Wind
- d) Ocean currents

Which action helps fight deforestation?
- a) Cutting down more trees
- b) Replanting forests ■
- c) Using plastic products
- d) Building highways

What is a simple way to reduce your carbon footprint?
- a) Drive alone every day
- b) Eat less meat ■
- c) Keep appliances plugged in
- d) Waste more water

How can students help the environment?
- a) Use disposable plastics
- b) Recycle and plant trees ■
- c) Waste paper
- d) Ignore pollution

ENVIRONMENTAL LAWS & GLOBAL INITIATIVES

What is the name of the international agreement aimed at reducing carbon emissions?
- a) Kyoto Protocol
- b) Paris Agreement ■
- c) Earth Summit
- d) Green Deal

Which law protects endangered species in the U.S.?
- a) Clean Air Act
- b) Endangered Species Act ■
- c) Wildlife Protection Act
- d) Marine Conservation Act

What global event raises awareness about climate change?
- a) Black Friday
- b) Earth Hour ■
- c) Cyber Monday
- d) April Fools' Day

What organization monitors global climate change?
- a) NASA
- b) United Nations
- c) IPCC ■
- d) Greenpeace

What is the main goal of the Clean Water Act?
- a) Prevent ocean pollution ■
- b) Regulate oil drilling
- c) Encourage deforestation
- d) Reduce electricity use

RECYCLING & WASTE MANAGEMENT

What is the symbol for recycling?
- a) A leaf
- b) A green circle
- c) Three arrows forming a triangle ■
- d) A water droplet

What type of waste takes the longest to decompose?
- a) Paper
- b) Plastic ■
- c) Food scraps
- d) Wood

What does the term "zero waste" mean?
- a) Producing no trash ■
- b) Using less water
- c) Reducing food waste
- d) Throwing away plastic

What material is the easiest to recycle?
- a) Glass ■
- b) Plastic
- c) Styrofoam
- d) Fabric

What does "upcycling" mean?
- a) Recycling old materials
- b) Burning waste
- c) Turning waste into higher-value products ■
- d) Throwing trash in a landfill

SUSTAINABLE AGRICULTURE & FOOD SYSTEMS

What is the main cause of deforestation?
- a) Urbanization
- b) Agriculture ■
- c) Climate change
- d) Air pollution

What is an example of sustainable farming?
- a) Monoculture farming
- b) Overfishing
- c) Crop rotation ■
- d) Deforestation

What is the environmental impact of eating less meat?
- a) Increases carbon emissions
- b) Reduces greenhouse gases ■
- c) Causes pollution
- d) Uses more land

What is "organic farming"?
- a) Farming without chemicals ■
- b) Large-scale farming
- c) Growing plants with plastic
- d) Raising animals in small cages

What is a "food desert"?
- a) A place with no restaurants
- b) An area with limited access to fresh food ■
- c) A region with too much food
- d) A type of farming method

FRESHWATER CONSERVATION & WATER POLLUTION

What is the biggest source of freshwater pollution?
- a) Oil spills
- b) Industrial waste ■
- c) Ocean currents
- d) Volcanoes

What percentage of the Earth's water is freshwater?
- a) 3% ■
- b) 10%
- c) 30%
- d) 50%

What is the largest source of freshwater on Earth?
- a) Rivers
- b) Lakes
- c) Glaciers and ice caps ■
- d) Underground wells

What human activity uses the most water?
- a) Drinking
- b) Agriculture ■
- c) Showering
- d) Car washing

What is an example of a water conservation practice?
- a) Taking long showers
- b) Fixing leaks ■
- c) Washing clothes daily
- d) Watering lawns every day

DEFORESTATION & REFORESTATION

What happens when too many trees are cut down?
- a) More oxygen is produced
- b) Carbon dioxide levels increase ■
- c) The ozone layer thickens
- d) Wildlife populations grow

What is one way to restore forests?
- a) Build highways
- b) Reforestation ■
- c) Increase logging
- d) Burn trees for energy

How does deforestation affect wildlife?
- a) Provides more homes
- b) Destroys habitats
- c) Improves biodiversity
- d) Helps animals migrate

What type of forest is most at risk?
- a) Boreal forests
- b) Rainforests ■
- c) Mangrove forests
- d) Temperate forests

What is a benefit of planting trees?
- a) Increases pollution
- b) Absorbs carbon dioxide ■
- c) Reduces soil quality
- d) Decreases biodiversity

EXTREME WEATHER & NATURAL DISASTERS

What type of disaster is worsened by climate change?
- a) Earthquakes
- b) Hurricanes ■
- c) Volcanic eruptions
- d) Solar storms

What is the leading cause of rising sea levels?
- a) Rainfall
- b) Melting ice caps ■
- c) Earthquakes
- d) Oil spills

What is a heat wave?
- a) A rise in sea temperature
- b) A prolonged period of extreme heat ■
- c) A type of ocean current
- d) A cold winter

What happens when permafrost melts?
- a) Releases greenhouse gases ■
- b) Produces fresh oxygen
- c) Increases ice formation
- d) Strengthens the ozone layer

What can help reduce damage from floods?
- a) Cutting down trees
- b) Restoring wetlands ■
- c) Increasing urban development
- d) Removing riverbanks

MISCELLANEOUS ENVIRONMENTAL KNOWLEDGE

What is a carbon footprint?
- a) The weight of carbon in the air
- b) The amount of carbon dioxide emissions caused by human activities ■
- c) A type of pollution
- d) A natural process

What is the name of the movement that promotes plant-based eating to help the environment?
- a) Zero Waste
- b) Meatless Monday ■
- c) Fast Food Nation
- d) Carbon Warriors

What does an environmental activist do?
- a) Protects the environment ■
- b) Cuts down trees
- c) Increases pollution
- d) Promotes fossil fuels

What is the "Great Pacific Garbage Patch"?
- a) A recycling center
- b) A large area of floating plastic waste ■
- c) A coral reef
- d) A shipping route

What does the ozone layer protect us from?
- a) Hurricanes
- b) Ultraviolet (UV) radiation ■
- c) Carbon dioxide
- d) Air pollution

FINAL QUESTIONS

What is an example of a non-renewable resource?
- a) Solar energy
- b) Wind energy
- c) Coal ■
- d) Hydropower

What is an effect of ocean acidification?
- a) Coral bleaching ■
- b) More fish
- c) Clearer waters
- d) Increase in oxygen

What percentage of Earth's biodiversity is found in the ocean?
- a) 20%
- b) 50%
- c) 80% ■
- d) 90%

What is an example of an invasive species?
- a) Polar bears in the Arctic
- b) Zebra mussels in the Great Lakes ■
- c) Kangaroos in Australia
- d) Oak trees in forests

What is the best way to educate people about environmental issues?
- a) Watching TV
- b) Environmental campaigns ■
- c) Ignoring the problem
- d) Cutting down more trees

What is a major cause of soil erosion?
- a) Planting trees
- b) Deforestation ■
- c) Wind turbines
- d) Recycling

What is the largest living structure on Earth?
- a) Amazon Rainforest
- b) Great Barrier Reef ■
- c) Grand Canyon
- d) Mount Everest

Which country has the most eco-friendly policies?
- a) USA
- b) Sweden ■
- c) China
- d) India

What is one way to promote sustainability in schools?
- a) Using disposable plastics
- b) Recycling programs ■
- c) Wasting food
- d) Ignoring environmental issues

What is the goal of Earth Day?
- a) Celebrate technology
- b) Promote environmental awareness ■
- c) Increase deforestation
- d) Burn fossil fuels

YOUR VOICE MATTERS!

Now that you've explored the amazing world of Earth facts and tested your knowledge with trivia, we hope this book has left you feeling more inspired and informed. But learning is just the first step—taking action is what truly makes a difference!

● **What can you do next?**
- Share what you've learned with friends and family.
- Take small eco-friendly actions like recycling, reducing waste, and conserving energy.

- Spread awareness by leaving a review and sharing your thoughts—your feedback helps inspire others to join the movement!

💬 We'd love to hear from you! What was your favorite fact? Which trivia question surprised you the most? Leave a comment, share your ideas, and let's keep the conversation going. Together, we can make every day Earth Day! 🌱 🖤

Made in United States
North Haven, CT
11 April 2025